Of Time and Tide

(An Alternative View of the Universe)

Malleus Maleficarum, Jr.

DENVER, COLORADO

APPRECIATIONS

Of course, an appreciation goes to the many giants of science and learning: Einstein, Hawking, Kepler, Copernicus, Galileo, Neil deGrasse Tyson, Socrates, Freud, Jung, and to the many unnamed whose work and influence are the foundations of these ideas. With equal importance, I must note my appreciation of the men, women, and students who struggle daily to learn and grow and increase their knowledge and help others against the oppressive forces of superstition and dogmatic thinking and mindless trivia. Much appreciation goes to my editor, Alanna Boutin, for taking my illiterate musings and shaping them into intelligible manuscript minimizing my many errors. And a great big "thank you" to Lois Butler for the seascape painting on the book cover.

QUOTATIONS

Tai T'ung (thirteenth century): "Were I to wait for perfection, my book would never be finished."

"I know that most men, including those at ease with problems of the greatest complexity, can seldom accept even the simplest and most obvious truth if it be such as would oblige them to admit the falsity of conclusions which they have delighted in explaining to colleagues, which they have proudly taught to others, and which they have woven, thread by thread, into the fabric of their lives." — Tolstoy

Contents

PART I:

PROPOSITIONS AND CONCLUSIONS

Chapter 1
TIME

WE ALL KNOW what time is. We all have watches and clocks and calendars. We rule our lives by time, by schedules. Our daily events are run by our clocks; waking up to our alarms; hurrying off for our daily routine of school or work; we arrive at our job or school and start our work or class at a specific time, and then, at another scheduled time, we pause our work for a coffee break or end one class to move to another; we have our lunchtime by the clock and our workday/school day ends by the clock. We go home where we are "off the clock," but our lives are still controlled by time, a driven routine—eating dinner at the same time, watching our favorite TV show at a certain time, e-mailing and chatting on the phone or computer, and going to bed at our normal time. On days off we spend our free time, but still schedule times to meet our friends, shop within store hours, and dine within restaurant hours. We may not be ruled by the clock in our off time, but are guided by various time factors and habits. We all mark our calendars with scheduled appointments for the dentist and special activities outside our daily routine. We have our schedule planners in booklet form or on our computers or mobile electronic devices to remind us of special days that occur annually . . . birthdays, anniversaries, etc. We are so immersed in the clock and calendar, in the present minute, in the daily minutiae of our lives, that we may forget what the clock and calendar represent. The clock is a measure of the rotation of the earth on its axis; the calendar is a measure of the earth's movement around the sun. In other words, the clock and the calendar are our measure of the earth's travel in our little corner of the universe, or as Einstein phrased it, through space-time, and since we are passengers on this earth, the clock and calendar measure our passage through space-time.

Let us, for a moment, look back into the origins of the concept of time. The DNA of most of earth's creatures is built around time. Most animals are either diurnal or nocturnal, spending the majority of their active hours either during the day or at night. The bodies of the different creatures have evolved to take advantage of their particular niche in the environment. Nocturnal creatures often have large eyes to be able to see well in the dark or a heightened sense of smell or sound. Migrations are timed to coincide with seasonal patterns. The great herd migrations in Africa are timed for rainy seasons, and the flourishing of the savanna and the predators follow the herds. Bird migrations are timed with the warming northern climates when there is an abundance of food and nesting sites. Rutting is timed with gestation to provide birthing when there is maximum food available for the best survival chances of the offspring. In the African savanna, this is when the rains come; in the temperate zones, it is spring. The remarkable monarch butterfly flies north following the warming weather and blooming flowers over four or five short generations, and then returns to the Mexican forest in one generation in a migration not yet understood by science. Temperate zone plants grow and flower with the warmer weather; bees and insects pollinate the flowers and propagate the plant species. The grazing animals and birds often eat the flowers, fruits, and grasses, and spread the seeds. Many of the earth's creatures are fitted into their niche, their daily routine determined by the annual rotation of the earth around the sun and the earth's axial wobble. We cannot make any absolute assertions about the influences of time on the creatures in the age of the dinosaur, but the fact that we find nesting sites with multiple nests indicate similar behavior for rutting, gestation, and birthing. A couple sites of massive jumbled single species kill off suggest a river crossing gone awry and supports the idea of seasonal migrations, inferring similar behavior in the dinosaurs then as in the mammals of today.

It is generally accepted that the common primate ancestor of the great apes and mankind arose in the tropical forest of Africa. Because of the relative consistency and abundance of food in the tropical forest, mating and birthing continue throughout the year, although there may be some seasonal variations. This strongly suggests that the overwhelming concern of our early ancestors was the clock; that our ancestors responded to day and night, hunting for food by day and bedding safely at night so as not to become food, not the calendar since the seasons brought little variation in their behavior. It is an unresolved question as to why our primate ancestors came out of the arboreal existence to become the biped ground dwellers.

The generally accepted theory has long been that as the climate dried out and the forest thinned out so our ancestors were forced to adapt or die. Some recent pollen analysis from the sediment layer of the remains of these earliest hominids indicated that the forest was still very much intact at this time. This led to a new theory for bipedalism, that those males able to carry back the most food got the most sex, therefore, began walking upright to carry food in both arms, and this trait was passed on to the offspring of these males who were granted the most sex. This theory of food for sex is still current today in the dinner date or in the more direct approach among the impoverished such as Russia in the 1980s during their economic collapse and the arrival of American fast-food chains when girls were reputed to exchange sex for hamburgers.

The food for sex idea does not seem to have any ready examples in either our closest primate relatives or known primitive cultures. The three main steps of evolution are isolation, mutation, and radiation. With groups that are isolated, often by environmental catastrophe, mutation occurs in the small group and when it is successful, the new and improved creatures thrive and spread. With the territoriality and rudimentary hunting we see in our nearest primate cousins, it may well have been a hunting culture that developed and drove bipedalism for tool use (crude weapons) and organization. In a self-sustaining cycle, the higher protein, the faster brain development, better tools (weapons), and better diet. While the sites of earliest hominids produce no tools, we are not talking about sophisticated objects but crudely fashioned sticks for spears and heavier limbs or rocks as clubs. It is noted that our ancestors scavenged the kill of other animals, but that too is a form of hunting. Usurping the kills of others takes organization and weapons (tools) to drive away the killers or other scavengers. But our discussion is not of our evolution, but the relation of man to time. The clock remained preeminent as our primitive ancestors emerged and matured. An understanding of seasonality probably emerged for the hunting of migratory animals. When the hominid migration reached the temperate zones, the cycles of weather blended with the behavior of animals and the changing flora, and thus, a concept of an annual cycle was incorporated into mankind's understanding.

About 10,000 years ago as man changed from hunter-gatherer where the annual seasons were used to predict migrations and harvests, to the farmer, a more definitive measure of the seasons became even more important, for it now included planting and harvesting. Farming allowed for denser populations, the formation of permanent cities, and the leisure time for the formalization of knowledge. Among the immediate and most pressing issues was the measure of time. As calendars were created, they took on a magical influence over man for their predictive value, not only planting times but tides, eclipses, and other mysterious events. Even in our modern, supposedly sophisticated era, the calendar is associated with predictive values and occasionally causes anxiety for a catastrophic event at a certain date at the turn of the century or the end of the Mayan calendar in 2012. Stonehenge was a gigantic calendar marking the solstices for farming and also a focus of religion and mysticism. Many ancient temples appear to have some religious significance through the predictive values of understanding the solstices and seasonal changes of the earth.

As man's knowledge increased so did his mastery over time: a single rotation of the earth was divided into more than day and night. In antiquity, sundials were used as well as a variety of water clocks dating as far back as 4000 BC. Greek and Roman civilizations are thought to have had some rare mechanical time devices. As early as the thirteenth century, European mechanical time devices began to appear. In the fifteenth and sixteenth centuries, spring-powered clocks enabled greater accuracy. Clocks gave way from the single-arm device with hours divided into 15-minute periods, into clocks with two hands measuring minutes as well as hours. Soon the third hand was added measuring seconds. The pendulum clocks of the seventeenth century provided another drive mechanism with excellent accuracy. Not only did the accuracy of the clocks improve, but some incorporated lunar phases, tying together the larger view of time.

We are agreed that the clock measures the earth's rotation on its axis and the calendar measures earth's travel around the sun. Time has always been the measure of the earth's travel through space, and as man is of this earth, so too it is a measure of our trip through space (or as Einstein phrased it, through space-time). Because of the earth's consistency in movement during our lifetime, we have come to assume that our passing through space-time, or as we think of it, as the passing of time as always being constant and consistent. We live our life driven on this time scale of ours, miles per hour, or gallons per minute, or vitamins per day. We are so embedded in our concept of time, we often forget it is a measure of our passing through space-time.

Einstein said that if we move through space-time faster, less time would pass. We are so used to thinking of time as absolute, that it sounds strange, at first, but it is really an easy concept to understand: if the earth were spinning twice as fast as it is now, then our day would be half as long as it is now. So time varies by the speed we are passing through space-time. Einstein's example of this is the twin paradox. The twin paradox is briefly that one twin flies off into space at a high speed and returns at a high speed several years later. The twin who was traveling faster in the spaceship is younger than the twin left behind on earth. At first, this may sound paradoxical or mysterious, but if we remember that time is the measure of our passing through the universe (space-time), then it is obvious that the faster we go on our journey, the less time it takes.

I think a better example of the speed-time relation was given by Nigel Calder in his book *Einstein's Universe*. Two spaceships are flying past our earth at half the speed of light. The spaceships are staying 50 light nanoseconds apart. To check that the ships are in the correct position, one ship bounces a laser beam off the other ship, a total of 100 nanoseconds. To an observer on earth, this action took 15 percent longer, or 115 nanoseconds. This more dramatically illustrates that time is relative, that a single action can be timed differently depending on the relative speed of the observer.

Being within the constancy of time and speed here on earth, we may often forget that time is relative to speed, but we need to keep that in mind as we further examine time and ourselves and our relation to the universe. While we can see the relation between time and speed in Einstein's relativity without much trouble, another related proposition of his theory seems to show even greater intuition which is that massive rotating bodies in space would drag time with them in the direction of rotation.

Up to the time of Einstein's Theory of Relativity in the first couple of decades of the twentieth century, the focus of timepieces was simply a better and more accurate way to measure the movements of the earth, but with the advent of relativity, new measures of time were needed to test his theory.

Einstein chose the speed of light in a vacuum as the constant in the universe from which to work, but there remain problems when we use this for practical measurements in the dynamic universe where there is no fixed distance for measurement, nor absolute time for reference. To take our measurement of time to an abstraction necessary for science and technology, we need to get out of the variations in the earth's rotation and orbit and out of the speed variances in Einstein's relativity. Enter the atomic clock. The first accurate atomic clocks were premised on the resonant frequency of Cesium 133. Since the first experimental atomic clocks in the 1950s and 1960s, accuracy has continued to improve until, with the Quantum Logic Clock, accuracy, can be predicted with an error of less than one second over a billion years. Now rather than dividing the day into hours and the hours into minutes and the minutes into seconds, a second is defined as 9,192,631,770 cycles of radiation from the Cesium 133 between the two states of energy. Time is now standardized.

When Einstein first proposed in relativity that large rotating bodies in space dragged time with them, there was no means of checking that assertion. After the invention of the first atomic clocks which created a measure of time independent of the earth's movement and the invention of air travel, the stage was set to test that portion of relativity. In 1971 came the famous Hafele-Keating experiment where four Cesium clocks were flown around the world twice on commercial airliners, once east, and once west. The experiment confirmed not only that the higher speed traveled in the airliners showed that less time passed than at the stationary reference clock, but that the time variation was dependent on direction of travel, with or against the earth's rotation. That experiment confirmed that earth drags time with it. As technology progressed and the military created the Global Positioning System, a correction had to be included for the time shift caused by the earth's rotation. Since it is accepted science now that the earth drags time with it, we now have very accurate clocks that are independent of the earth's movements, so let us move on.

How does the human personality change a lightbulb? In evolution, an animal is the sum total of its instincts: food, fight, flight, rest, procreation. As creatures evolved and their behaviors became more complex, the instinct toward self-preservation became rudimentary self-awareness. Evolution pushes creatures toward more complex behaviors and learning through greater intelligence. In the higher primates and hominids, this intelligence reached new levels, yet even in modern man, most of our complex behaviors can still be traced back to our basic instincts.

The instinct of self-preservation evolved into self-awareness in the higher primates, and self-awareness in mankind evolved into self-importance. We are all the centers of our own world and create any illusion to keep it so. On the greater scale we have created anthropocentrism, which is the view that mankind, and by extension, ourselves, are the center of the universe. This anthropocentric view has always existed in our thinking and has always provided a great obstacle in science which tries for objectivity but often does so under the preconceived anthropocentric view. So how does the human personality change a lightbulb? It holds the bulb up to the light socket and waits for the world to revolve around it.

Anthropocentrism has, as we might expect, provided the wrong perspective about time. In our overblown sense of self-importance, we have viewed time as a river passing us or carrying us along. Time travel was always conceived as jumping out of the river at the present time and re-entering it upstream in the past or downstream in the future. Even science often has this river view of time, but it is very simple if we simply remember what time is. Clocks are the measure of the earth's axial rotation, and the calendar is the measure of each earth rotation around the sun. Thus, time is the measure of the earth, and by extension, our travels through the universe. We are all time travelers here on earth traveling through the universe. If there is a water analogy to be used in time, it is that time is like an ocean and that we, on the good ship *Earth,* are traveling through it. We still say the sun rises, even though we now understand it is not the sun moving around the earth, but the rotation of the earth on its axis that causes the day and night for us. In a like manner, we can still say time passes . . . but we need to remember, it is not time passing, but rather, we, who are passing through time.

Chapter 2
TIDE

TIDES HAVE PLAYED important roles in the development of life here on earth. It is thought the tidal zones are where plant life first evolved from living only in the oceans to living outside of the protective immersion in the seas. While plant life soon covered the earth's land masses in a proliferation of forms and diversity, plants did not lose the need for water, usually accessed through a root system. These root systems and the water absorption by plants greatly reduce the erosion which marked the early earth. With the plants utilizing this new growing area on the land, animals soon began evolving from these same tidal pools, to survive outside of the oceans and to take advantage of this new food source.

Hundreds of millions of years after the transition of life from ocean to land, early hominid hunter-gatherers would use these tidal zones as an easy source of food.

Early man noted the relation of the tide and the phases of the moon. While day and night were the first measures of time, the moon phases may well have been the second. The biggest problem with the lunar calendar is its 28-day cycle does not work evenly into the earth's rotation around the sun; nonetheless it is a very important marker spaced between days and years, and useful tool in measuring and communicating this time span, while other markers could be used for seasonal changes, temperature, length of days, flowering and fruiting of plants, and migration or behavior of animals. About 10,000 years ago, when man began to change from hunter-gatherers to farming and larger more permanent settlements, then the calendar of importance became the year for planting, but prior to that time, the moon ranked high for primitive man as a measure of time.

The relationship between tides and phases of the moon appears as far back in mankind as we have knowledge of early man. Clam diggers today usually understand that the tide is related to stages of the moon, but are more likely to consult a published tide table or the Internet than consult this archaic tide table: the moon. I heard a story in the 1960s that a bucket of clams in a basement some 200 miles from the ocean, where they received no visual clues from the moon, behaved as if it were low tide when the tide would have reached them there. At the time I thought it was an interesting story .

.. however, not so interesting at the time that I bothered to find out what kind of clams, nor inquire how clams behave at low tide. While I can find no reference to this story searching Google, there is, however, a study posted online that we can use for our purpose. The study is "Discrimination Among Wave-Generated Sounds by a Swash-Riding Clam" by Olaf Ellers, department of zoology of Duke University at Durham, North Carolina. The *Donax variabilis* clams come up from the sand and jump into waves during incoming tide, high tide, and outgoing tides. Ellers's study performed with *Donax variabilis* clams in an aquarium is primarily to determine what stimulus causes this behavior. It is one section of the report which I find most relevant to the bucket of clams in the basement story. Both on the beach and in the aquarium the clams do not jump within 1 to 2 hours of low tide. He also notes that this is determined by tidal time, not by time of day, and there was no water level change in the aquarium, or other visual stimulus. Ellers's study is closely in accord with our bucket of clams in the basement story.

Most everyone, scientist and laymen alike, are in agreement that the tides on earth are caused by the gravitational forces of primarily the moon and secondarily, the sun. This being so, when we are considering ocean tides, we need to address gravity.

Chapter 3
GRAVITY

WITH OUR ANTHROPOCENTRICITY, prior to the discovery that the earth revolved around the sun, the geocentric view which dominated was that the earth was the center of everything, that we lived on a flat earth, and if we reached the ends of the earth we would simply fall off into nothingness, much like falling off a cliff into a bottomless valley. The sky was believed to be a multilayer sphere rotating around us, the sun in one layer, the moon in a second, the planets in others, and the stars in yet another layer to explain the variations in movement. No one really seemed to think of the concept of gravity. Up was up and down was down, and any damn fool knew when something fell, it fell down, not up. Even after the idea that the earth was round began to gain acceptance the geocentric view still prevailed.

The origin of the concept of gravity came with the

rude awakening that the earth is not the center of the universe. As Johannes Kepler, Nicolaus Copernicus, and Galileo Galilei developed insight into the heliocentric view of the solar system with the sun as the center and the planets orbiting around it, they needed a reason for these planets to stay in orbit, rather than flying off into space. The only solution they could come up with is that there is a physical attraction between objects in space depending on mass. The brilliant Sir Isaac Newton provided the mathematics for this force and called it gravity, although he did not seem satisfied with that concept. Of course, the human personality gets pissed off when you tell it that everything does not revolve around it, so they gave Galileo the choice of recanting his view on the solar system and live . . . or take his heliocentric view to an early grave. Galileo, faced with those choices, recanted and repented and although he wasn't allowed to play with his telescope or go out after dark any more, he still managed to continue experimenting and wrote "Discourses and Mathematical Demonstrations Relating to Two New Sciences" with important principles in strength of materials and the motion of objects, which survived even after great repression.

In the first couple of decades of the twentieth century, Edwin Hubble and others refined the heliocentric view so that it still held for the solar system, but placed this solar system not at the center of the universe, but rather insignificantly in the Milky Way on the edge of this galaxy amid millions of other stars, and later, the Milky Way was found to be simply a common galaxy among the millions of other galaxies in this universe. The human personality should have suffered greatly from these revelations, but fortunately has the great ability to ignore facts and continue its anthropocentric delusions, even showing humans zipping around the universe righting the wrongs of those bad aliens.

When Einstein published his *Theory Of Special Relativity* in 1905 and in 1916, his *Theory Of General Relativity*, he proposed that large bodies in space bend space-time and that the planets traveling around the sun actually travel in a straight line through space-time, but that space-time is bent by the mass of the sun so there is no physical attraction between the planets and the sun. While much of the *Theory Of Relativity* has been widely and readily accepted in the scientific community and largely incorporated into current scientific thinking, there are exceptions as researchers try to prove or disprove certain aspects of relativity. We have seen that the speed we travel affects time, and we know that the earth drags time with it. This is now accepted and fully incorporated into the practical aspects of our lives. The bending of space-time by large masses in space, although famously corroborated during the full solar eclipses by the bending of starlight near the sun, is still questioned by many scientists. Einstein used this bending of space-time as the cause of planets rotating around the sun, which, as you remember, was the original concept used to explain why the planets stayed in orbit around the sun. While the curvature of space-time is often taught, most folks, scientists included, are reluctant to give up gravity as an attractive force between masses. The supposed attraction between masses is reputed to be caused by a particle called the graviton. The search for the elusive graviton has been going on for a century without success, and continues today.

Gravity's big brass balls! In one experiment designed not to find the graviton but simply to prove the concept of attraction between massive objects, a university designed a pair of very heavy brass pendulums suspended from a precisely leveled rod in a ridged frame set on a precisely leveled floor. The pendulums were set in motion to overcome inertia. The reasoning was that these two very heavy moving objects would gravitate toward each other (however slightly) from the attractive forces of gravity. It was an excellent, well planned and executed experiment, but the pendulums did not move toward each other. While it does not disprove the attractive forces between massive objects, it did fail to affirm it.

If I were to tell you there is no gravity, you might very well hold this book over an open garbage can and releasing your grip, watch it fall helplessly into the rubbish, proving to yourself that gravity is, indeed, alive and well where you are, but when a man of Einstein's intellect tells you there is no gravity then we must take a closer look. What Einstein meant when he said there is no gravity is that the planets are simply traveling in a straight line through curved space. So if there is no gravity why aren't we all jumping over mountains like we had shoes made of flubber? Flubber is the gravity-defying substance from the 1997 movie *Flubber*.

As Dr. Michio Kaku noted recently on a television series Einstein believed that space-time pushes against the earth and large bodies in space, which causes the effect we call gravity (which keeps things from just falling off the earth). In essence, there are two theories of gravity: Einstein's push theory that space-time pushes against the earth and everyone else's pull theory—that holds there is an attractive force pulling objects toward each other. It needs to be understood that whichever theory is correct it doesn't change the calculation for gravity anymore than discovering that the earth rotated on its axis, rather than the sun's rotation around the earth changed the length of the day. There we have the present contest: Einstein verses just about everyone else. In this paper, we are standing with Einstein. The problem is that the good doctor did not explain the mechanics of how his push theory of gravity was accomplished. Perhaps his feeling was, "Hey, I have more important things I want to work on, so you can figure it out for yourself," or perhaps he did not quite understand the mechanics of this force. In either case, we'll work on it here.

Our first clue we'll call "Skinny Dipping for Einstein." It is well-known by the younger crowd that skinny-dipping reveals many secrets and answers many of our questions. The next time we are skinny-dipping at an ocean shore at night and our personal questions have been addressed, we might note that when the moon is above us, the tide is near ebb. No one disputes the moon is the primary cause of the tidal force on earth as the tides are sequenced with the orbiting of the moon around the earth, rather than the rotation of the earth on its axis, but it appears by the tide that gravity is pushing, rather than pulling. It would seem that if the pull theory of gravity was correct, that when the moon is above us, the tide should be high; but, in fact, it is just the opposite. If we were to express this as a scientific principle in the technical language of science it would be: "Where da moon is, da ocean ain't." Of course, those who subscribe to the pull theory of gravity have noted this too and have explained it in one of two ways . . . One is that gravity pulls from the edges of the earth and the moon, rather than the centers. The other explanation is that due to the rotation of the earth on its axis, the oceans slosh around and end up in the inverse position of what one would expect.

Now let us put together these ideas noted above and see what we come up with. If time is dragged with the rotation of the earth and large masses in space bend space-time and space-time pushes against us to create what we call gravity, then, it is reasonable to assume that space-time is a physical property of the universe. It is said that Einstein thought of time as another dimension, but let's try it out as a physical property and see how it would work here, in our little corner of the universe. While Einstein phrased it as space-time pushing against matter, I prefer to think of the situation as just the opposite; that is, that matter forces itself into space-time.

In our solar system, with which we are most familiar, our massive sun compresses space-time. This compression is highest near the sun and diminishes the further we move from the sun. In the theory of relativity, this is the "bending of space-time." The planets travel around the sun according to their size and speed in the compressions of space-time. The moon travels around the earth in the compression made in space-time by the earth, but the moon too compresses the space-time, and as the moon travels around the earth, it adds additional compression to space-time between itself and the earth, and causes the tidal forces here on earth. So the moon's additional compression of space-time is pushing against the earth's surface and pushing the waters of the oceans away, causing low tide under the moon, and the displaced waters that are pushed out from under the moon are causing the high tides in the lower compression areas. (To help explain the tides we can easily see what we will call the "balloon effect." Within the closed system of our ocean of space-time we get a balloon effect where we squeeze harder in one place, and thus, the balloon pops out in another place. Here on earth as the moon adds additional compression of space-time between itself and the earth, it also squeezes space-time out of the highest compression area in a ballooning effect contributing to the high tides). Of course, the big daddy, our sun, not only provides the space-time compression that our earth travels in (bending of space-time in relativity), but also adds additional compression to between itself and the earth, affecting the tides on earth by its relative position to the moon. The moon's influence on our tide is greater due to its close proximity than the influence by the sun at a much-greater distance. The sun's affect on the tide, whether working with the moon or against the moon, has always been measured easily by phases of the moon.

Last, but equally important, remember our bucket of clams in the basement or the notes from Ellers's study that the clams respond to low tide, although in both cases there was no change in the water level and there were no visual clues. We can now understand that the clams are not responding to the tide, but rather to the compression made in space-time here on earth by the moon.

Einstein used to use the term *space-time*, but since space just denotes a void, let's just say that the part of universe that is not occupied by matter, the great voids between the stars and galaxies, is full of physical time. This fits nicely with our understanding of our concept of time which originated from the earth's travels spinning on its axis as well as its journey around the sun. Massive objects in space bend physical time (previously space-time). This is another way that time interacts with matter. If time is a physical property, then we need to try to determine its physical form. We are used to dealing with physical things we cannot see, like air which was known because we breathe it and because of the wind and even though ancient civilizations did not understand air scientifically, they were able to acknowledge it and deal with it. In a like manner we can take what we know of time and propose a physical definition. We know that massive objects in space bend (distort) time-space, and that the earth drags time-space with it. Since objects in space are nominally spherical in shape and that these are flexible (distorted by massive objects), let us make them tiny time bubbles. We can call them tymeatrons. Massive objects force themselves into this mass of time bubbles compressing the time bubbles. The nearer the time bubbles to the solid mass, the more the compression, and this creates the "bending of time and space" around massive bodies. The rotation of the large body drags the time bubbles with the most movement at the equator, causing a disruption or current in time-space and the orbital plane for subject bodies.

Chapter 4

TYMEATRONS
(time bubbles)

I KNOW A lot of you are thinking *tymeatron!* Really? Is that what we need . . . another mystical likely mythical unfound particle? Just because we have not been able to find the graviton, do we have to have a new particle and a new search for the needle in a haystack? If we are to continue the haystack analogy, we always seem to be searching for the needle, but no one seems to notice the hay.

It has been known to Western science since 1660 that air was required for the transmission of sound (ringing bells make no sound in a vacuum jar). As we began to understand the great void of space, it was assumed by some scientists that, like the bell needing air to transmit sound, light too must have a transmitter. Hence, a property called luminiferous ether was deduced to fill the great voids of the known universe. By the mid-1900s, this concept had faded from science to reside mainly in some science fiction novels. By proposing that the great voids of this universe are filled with physical space-time tymeatrons, these tiny time bubbles—we have to decide how light travels through them. Do photons of light pass through these time bubbles, using them as conduits, or do they slip between the bubbles? The question is not difficult to answer. Since matter is mostly exclusive of time, that is, it pushes into time, then so too is it likely that photons of energy (another form of matter) are mostly exclusive of the tymeatrons . . . That is, that photons of energy push between the time bubbles. I think there may well be many experiments in the future to better describe the tymeatron, but there is already an experiment from the past which we will discuss now.

As we know, energy is another form of matter. Since matter intrudes into time-space, energy too must intrude into time-space. Light, being a photon of energy, would not travel through the time-space bubbles, but between them. One way to observe the interaction between energy and time-space is the old double slot film box from the first half of the twentieth century. In a box with two gated slots, when one slot is opened, it exposes the film only under that slot. That is, the photons of energy (light) push the time-space bubbles out of the way and impact (expose the film) directly beneath the slot. If we do the same with the other slot, it is the same. The light exposes the film only directly under the slot. But when we open both slots at the same time, not only is the film exposed directly under the slots, as we would expect, but we find photons of light striking the film in places other than directly beneath the slots. While it wasn't understood at the time of the experiment, we can now understand what was happening. This is because of the interaction between the photons of energy and the tymeatrons. With one slot open, the photons simply push the time-space bubbles out of the way and expose the film directly under the slot, but with both slots open, some of the tymeatrons are pushed away from one slot with enough force into the path of the photons in the other slot that they redirect the photons of light to places on the film outside the area directly beneath the slot. In higher energy beams like lasers, they push the time-space bubbles out of their way with even greater force, so there is less diffusion. In the case of black holes, the intrusion into time-space is so great, it likely collapses the local time-space bubbles; they are flattened, and no light can travel between the bubbles.

In spite of the widespread acceptance of relativity, science has been very reluctant to give up the attractive force between masses. The search for the graviton goes on. Without this attraction force between objects in the universe, current theories of the origin and behavior of the universe fall apart. Sorry, but in the vernacular of the hippies, gravity no longer sucks ... It blows. Let us take a look at the universe without the G in mathematic models, meaning gravity, as an attraction between masses by first looking at our little corner of the universe: our solar system.

Chapter 5
THE SOLAR SYSTEM

ALL CULTURES HAVE creation stories explaining our existence, which usually include gods or mythical giant creatures. As curiosity and inquisitiveness turn to more methodical thinking, science was born, but still with the anthropocentric bias. As the heliocentric solar system became widely accepted, the next question was how it came into being. The religious and those with an anthropocentric view continue to believe, as before, in whatever creation story they've found acceptable, no matter how the rest of the universe is arranged, but for those trying to use rigorous scientific methodology, they have had to use the scientific method and mathematics, but so far, there have been no satisfying specific theories.

When I was a boy, still well after Einstein and relativity, one theory for the creation of our solar system was that another sun (star) came close to our sun and pulled the planets out of each other and some of them were left circling our sun. This, of course, uses the widely accepted pull theory of gravity, but there are many things unexplained in this theory, like how these two suns were close enough together to pull matter out of each other, but not close enough to form a permanent interaction, and obliviously, how these two massive balls primarily consisting of burning (nuclear fusion) hydrogen would yield up these planets of such diverse elemental structure.

A recent TV series described the creation of our solar system as a cloud of cosmic dust drifting through our galaxy, that, for some unknown reason, started to swirl, and as it swirled, gravity caused the center to collapse on itself until the pressure and heat became so intense it ignited into our sun and the rest of the solar dust swirled around until gravity formed it into the planets and moons and other space junk in our solar system. Our solar system has a gaseous hydrogen sun, the rocky planets, Mercury, Venus, Earth, the asteroid belt, Mars, and then back to gas: the gas giants, Jupiter, Saturn, Uranus, Neptune, then the Kuiper belt of rocks and ice. Even with an attractive force of gravity, this lacks logical continuity. Einstein's famous error was that the universe is static, neither expanding nor contracting, but since Edwin Hubble's famous red shift observations, no one has ever questioned that the universe is expanding. This current theory of the creation is filled with logical inconsistencies . . . First, how, in an expanding universe, even with the pull theory of gravity, why a cloud of cosmic dust, instead of dispersing in an expanding universe, would gravitate together . . . Why it would start rotating . . . Why the lighter hydrogen would gravitate to the center and build enough mass and pressure to begin the fusion process, while the heavier elements which would be most influenced by the pull theory of gravity, instead of gravitating to the center of this cosmic cloud and into the sun, instead, formed into rocky planets outside the center point? But enough! The questions are too many and answers too few. Perhaps there is a psychological demand for a creation; we certainly have a history of creation stories which have carried over into our science.

Interestingly enough, this same TV series provided much of the answer in a different segment. Our solar system was not created, but rather it evolved. Our sun and the planets of our solar system are in two different stages of cosmic evolution. Stars, as well as our sun, begin life as balls of hydrogen of such massive proportions that the intense pressure from pushing into time causes the fusion process. Fusion is the joining of two element nucleuses forming a new element. The larger the star, the hotter and faster it burns. When a sufficiently large star matures to iron, that is, when the elements have fused together so that a large portion of the star's elements is iron, the fusion process will come to an end, and that end is dramatic.

It is called a nova, where the star explodes and all the elements it has created are blown out into the surrounding area.

Like any explosion, the debris is all manner of sizes from planet-size lumps, to moon-size lumps, to rock-size debris, to dust: a full range of sizes. As the star goes through the maturing process and nears the end because of the difference in the weight of elements, there is a process of stratification so when the star explodes, different lumps of debris will be rich in different elements, depending on their position in the star. While there will be a full range of elements in any of the large debris, some elements will dominate. It is because of this stratification and the process ending at iron that we find so many iron-core planets.

When the debris from this star nova enters into the gravity (time compression) of another star, it assumes an orbit by its mass and speed. Perhaps some of the large and smaller debris had already established orbital relations with each other. The larger debris may establish an orbital relation with the new star to become planets; some of the smaller debris may become moons or rapidly (in cosmic time) be assimilated into the larger bodies by orbital intersect. While we may not know enough at this point about whether solar systems are primarily formed by the explosion of nearby stars or perhaps even orbital partners, I think it likely that some wandering planets join established systems and might well cause upset to the old order. Perhaps we have a situation similar to the earlier discredited theory of another star passing close enough to our sun to pull planets out of each other. Maybe 5 or 6 billion years ago our sun had a larger orbit partner, which, about 4.5 billion years ago, went through a nova and gifted us all or most of our solar system, including our own home: Earth.

How do we know that our solar system evolved through capture, rather than by creation? Well, aside from the overall logic explained above, there is the axial rotation of Venus, Uranus, and Pluto, which rotate clockwise in relation to the sun, while the rest of the planets rotate counterclockwise. In a created system, there is no viable explanation of how these two planets and Pluto would rotate in the opposite direction. While it may seem that the planets and moons in our solar system would have had to travel too great a distance to be captured, let us remember that 4.5 billion years ago, the Milky Way galaxy was much smaller and the stars were closer, and we do not have the history and relative positions of the star or stars which provided our planets and moons. With capture as the source of evolution of our solar system, and Venus likely one of the newer arrivals, we must be aware that when we acquire new sibling planets in the future, it may very well be disruptive to the present order of our little corner of the galaxy.

Saturn's moons and rings provide another great confirmation that our solar system was evolved, rather than created. The Cassini spacecraft has shown that one of Saturn's moons, Enceladus, is heating under the tidal stress and spouting geysers into space, creating a new ring. Mimas, Saturn's heavily cratered, innermost moon, seems to have created the Cassini division between Saturn's widest rings, which suggests it entered Saturn's (gravity) time-compressed area at a different speed than the debris that make up the rings, and accumulated them by collision and/or pushed them out of the division.

With this basic understanding of how our solar system evolved, let us move to the larger view: the known universe; its origin and how it is evolving.

Chapter 6
THE UNIVERSE

THE NEXT MAJOR area I'd like to address is the theory of the beginning of this universe. Concepts of this universe are inherently difficult as our perspective has changed from the philosophers of Greece, who thought the earth was the center of all existence, to our understanding now that our sun is only one of the stars in a number that is more than all the grains of sand on all the beaches of our earth, and that mankind's existence is but a blink of an eye in the life of the universe. That said, the present theory known as the "big bang theory" of the beginning and construction of the universe appears to lack a logical consistency—from nothing to something—starting out as a tiny point of energy blossoming into a hot, dense state from which stars condense is very questionable. There are a couple of big problems regarding the theory. While mathematically zero to anything can be performed with ease, philosophically, it is a sticking point. It is easy to accept transformations, such as water to steam or mass to energy, but nothing to something is lacking in our world outside of magic tricks. Another point is stars condensing out of the hot, dense state stretches credible thinking. In this theory, a tiny speck of enormous energy rapidly inflates into all the matter and energy in the universe. In spite of this rapid inflation and continued expansion in this initial burst, hydrogen atoms, somehow, manage to group together and collapse into fusion furnaces we call stars. This idea seems to run counter to logic and any similar situation where the whole is expanding while internally, some points are collapsing. If we accept Einstein's push theory of gravity verses the pull theory of gravity, not even the most flexible and adaptable mathematics will hold up.

In 1965, Dr. Arno Penzias and Dr. Robert Wilson discovered a low level of radiation thought to be a heat signature that was constant in all directions at all times of the day and night and all times of the year, thereby concluding it was a universe-wide event. This heat signature has been used as a supporting fact for the big bang theory and indeed, is very supportive of a high temperature event in the early universe, but we might explore another possible scenario for this universe's initial event and its construction.

The idea of black holes, objects of such intense mass

that emit no light, not surprisingly were only conceived in the twentieth century to explain distortions in gravity fields. Prior generations could only deal with the blazing stars that could be seen and the local light reflecting objects. Since the theory of black holes and now the wide acceptance of their existence, it has added another dimension to the universe, which, although not fully understood, seems to have begun to explain some of this universe's construction. It is generally accepted that our galaxy has a black hole at its center. Further, it has been discovered that black holes are at the center of some of the other galaxies. The present thinking about black holes is that they are the graveyard of stars; that is, that stars close to the center of the galaxy are "pulled" into the black hole, that the mass of the black hole increases, and then drags in more stars. While that is true to some extent, let us not forget that we are in an expanding universe, which means it is expanding in every aspect; that is, that the moon is gradually moving further from the earth, and the earth is moving further from the sun; thus, the galaxies are expanding and moving further apart as the universe continues its relentless expansion. Expansion does not preclude objects within the various localities interacting with each other. We have meteors here on earth. We have suns near the center of galaxies interacting with the black holes. We even see galaxies colliding with one another. What we don't see is the asteroid belt assimilating into a new planet . . . More likely, it is the remains of a former planet.

It is time here to discuss CDA (Cockroach Discover Assumption). When we stagger out to our kitchen in the morning to make our coffee and turn on the light finding a cockroach, we assume that there are more, even though we don't see them at that time. When a scientist discovers a cockroach in his kitchen, he squishes it, which takes care of the cockroach problem. While we assume there are more cockroaches, the scientist has to discover cockroaches in his kitchen several mornings before he can assume he has a problem. This is because the rigors of scientific proof require more evidence. When science makes a new discovery, it is assumed it is an exception until it is proven to be the norm, but what we should be assuming is that the new discovery is the norm until proven otherwise.

When we find a supermassive black hole at the center of our galaxy, we should assume the same in all organized galaxies until proven otherwise. In fact, if there is no physical attraction between stars to form galaxies, then, indeed, there has to be supermassive object in the center of each organized galaxy to compress the tymeatrons (space-time) enough to create a galaxy. Since galaxies are the building blocks of the visible universe, and black holes are the centers of the galaxies, it is likely that black holes played a much-greater role in the creation of the universe.

If that is correct, it further suggests that the black holes are the primary building blocks of this universe, not the balls of fusion we call suns or stars, and that black holes are not simply the sometimes graveyard of stars, but have a parental effect . . . That is, that stars erupted out of the black holes. It has recently been found that unlike our solar system which has planets traveling around the sun at different speeds as a result of our solar system's evolution, accumulating planets by capture, star speeds in galaxies are consistent with each other, which suggests they had a common beginning, so speed measurements of stars in galaxies appear to support this idea. Thus the star speed supports the idea of a common beginning of stars in a galaxy.

The recent discovery of "HOT dog" galaxies in the early universe also supports this idea of galaxies erupting out of black holes. Whether new galaxies are being un-leashed now, or if they all occurred in the early universe, we do not know. The crab nebula and the eagle nebula, both considered to be the birthplace of new stars, do not look to me to be gases condensing into stars but more like the aftermath of an explosion. New stars cannot erupt out of dead stars. The old star died from running out of mate-rial for fusion, and while some of these elements are still present when the star undergoes nova, it is silly to think the hydrogen in the aftermath could overcome the force of the explosion to reassemble as a new star. New stars can only be born of a black hole or of some, as yet unknown, parent.

Black holes are a great mystery because we know so little about them. Although we know they have extraordinary mass, we do not even know if they are all of same composition. If they are the birth mother of stars, then we must assume they had basic elements of the hydrogen atom . . . a proton and an electron, which is a neutron when the electron enters the nucleus, but since black holes have more mass than a neutron star, we have to assume there is even greater compression. As subnucleonic research continues, perhaps a better understanding of black holes will emerge. We have gone from our timepieces and beaches to black holes, and so I reach the end of my speculation; therefore, let us review.

PART II:

REVIEW

Chapter 7

SUMMARY, ADDITIONAL THOUGHTS, AND NOTES

VOYAGES OF DISCOVERY have actually always served two purposes: to discover what is, but also to discover what ain't. While these voyages were always looking to find something new, they also dispelled old myths. Columbus and the early European voyages in the 1500s attempting to prove the world was round were expected by many of the Europeans to fall off the ends of the earth. They found "new lands," which, arguably, in many ways, had equal or superior cultures and science, but which they conquered primarily with germ warfare which had a greater and more decimating effect than the later bizarre military campaigns, although the conquerors neither understood nor intended that viral attack at the time. Not only had the voyagers made discoveries, they also dispelled the myth of the earth being flat with ends or edges.

We started our voyage of discovery here on earth, with a few thoughts on time, pointing out that time is what we knew it was . . . a measure of the earth's rotation as measured by the clock and the earth's revolving around the sun as measured by the calendar. It is the measure of the earth's travels through space and time, or as Einstein phrased it, through space-time, and since we are passengers on the earth, it is the measure of our travels through space-time. Because our lives are so short, in cosmic terms, and the movements of the earth so constant during our life spans, we had come to think of time as absolute. As Einstein explained, time is not absolute. It is always relative to the speed of the viewer. An easy example is if the earth was spinning twice as fast, a day would be half as long. Einstein used the example of the twin paradox, where one twin flies off into space for years at high speed, and then returns to earth at high speed, and when he meets his twin on earth again, the high-speed traveler is much younger than the twin who stayed on earth. Nigel Calder used the example of the two passing spaceships traveling at half the speed of light reflecting a spacing laser one off the other and the time observed for this single act being different when timed by the spaceship occupants and a slower-moving observer on earth. We also noted that as our science has become more exacting, it was necessary to measure time without reliance on the movement of the earth; hence, time is now fixed, for scientific purposes, with atomic clocks and no longer dependent on the movement of the earth.

So, our first great misconception about time is that it is constant, because here on earth, it is constant within the limits of human perception. The second great misconception about time is our old nemesis: our anthropocentricism, our psychological disposition to place ourselves at the center of our world. We all have this bias in our psyche, but we must get outside this prejudiced view even to see simple facts. We know that time is premised as the measure of the passage of the earth and ourselves through time and space. Yet, when we think of time, we always view it as flowing past us, when it is so obvious it is just the opposite. It is we here on the spaceship earth who are passing through time and space. In the water analogy, time is more like an ocean we are passing through, rather than a river flowing by us. We must always keep up the guard of our deductive reasoning against this anthropocentricism menace of assuming it is we who are primal.

The last—and to me, one of the most amazing—propositions of Einstein's Theory of Relativity is that massive rotating bodies in space would drag time with them. This part of Einstein's theory was corroborated in 1971 by the famous Hafele-Keating experiment which also affirmed that time is relative to speed. This dragging of time by the earth is now established and is corrected for in the GPS.

In our voyage of discovery, we moved on from establishing our thoughts on time down to the beach, where, after perusing the young ladies frolicking in the surf, we noted that when the moon is above us, the tide is out. Noting "the moons" in this part of our voyage of discovery, we made two conclusions: first, that skinny-dipping is great, and second, what begs to be titled, The Second Law of Lunacy: "Where da moon is, da ocean ain't." The First Law of Lunacy would have to be the observations made by thousands of generations of our ancestors that the moon is directly correlated to the tide and that the phases of the moon determine how high the high tide will be and how low the low tide will be. Of course, today, we know that the tides are caused by the gravitational forces, primarily of the moon, and secondarily, by the sun and the relation of adding to each other's gravity or opposing each other's gravity as noted by the phases of the moon revealing that relation. Of course, the tilt of the earth changes the proximity of portions of the earth to the moon for the seasonal changes.

The next stop on our voyage took us to the debate over gravity. We talked about the origin of the concept of gravity beginning with the description of the heliocentric solar system; that is, the planets of our solar system circling around the sun, rather than the geocentric view of all the planets and stars revolving around the earth. There was no known force which would keep the planets circling the sun, so some of the great minds of yesteryear made the obvious assumption that there was a physical attraction between these large bodies, that it was directly proportional to the mass of the bodies and inversely proportional to the distance between them. After a rough beginning, the new force was called gravity and became rapidly accepted throughout science and by the vast majority of educated people who now had a scientific reason for why it was so easy to get down and so hard to get up.

This view of gravity stood unchallenged until 1913 when Einstein published his Theory of General Relativity in which he postulated that there was no gravity; that, in fact, large bodies in space like the sun, curved space, and the planets were falling in a straight line through curved space. Interestingly, both ideas are still being taught. Subnucleonic research progresses and the search continues for the graviton: the particle thought responsible for gravity. Although experiments here on earth have failed to show the gravitational pull, we see that between the sun and the planets many scientists are reluctant to give up that idea, and we find that G popping up in many of the calculations for the universe. Who can blame them? After all, that G figures in our whole space program. Without it, we could not have successfully launched satellites or gone to the moon and returned or landed the rovers on Mars. True, Einstein said there is no gravity. What he meant is there is no attraction between masses. He said our feeling of gravity here on earth is not the earth pulling us, but rather space-time pushing against us.

Essentially, what the gravity debate comes down to is: Does gravity pull, as most people believe, or does it push, as Einstein believed? Either way, it does not change the calculations for gravity any more than understanding it was not the sun going around the earth, but rather the earth revolving on its axis that caused the sun's apparent movement changed the calculations for the length of a day. Although Einstein did not explain the mechanics of how his push theory of gravity works, other than time-space pushing against us, we chose to side with this brilliant man and tackle the mechanics of gravity.

My admiration for Einstein knows no bounds. He certainly possessed not only one of the greatest minds of the twentieth century—but of any recorded person in all of human history. He is known, respected, and loved around the world. It is then with some uneasiness that we deviate a little from his thinking. Einstein is said to have thought that time was another dimension. At least two portions of his theory of relativity seem obviously to contradict that. If massive objects in space bend space-time and massive rotating objects in space drag time with their rotation, that strongly suggests that space-time is a physical property of the universe.

We assumed space-time as a physical property, and then tested it in the tiny part of the universe we know best . . . our own local solar system. First, if space-time can be bent, as Einstein described, then we know it is pliable. We also know that there is a high degree of exclusivity between matter and space-time. I prefer to think of matter as inserting itself into space-time and not so much bending it, as compressing it, with the compression being greatest near the object and decreasing as we move away . . . This compression of space-time being what is called gravity.

The mass of the earth pushes into space-time. This, then, is the gravity we have on earth. The earth's moon orbits the earth in the space-time compression area fitting its mass and speed. The moon also pushes into space-time, causing gravity on the moon. The increased compression of space-time between the earth and the moon as the moon travels around the earth pushes against the surface, both earth and the moon, but here on earth, the increased pressure causes the tides, and it is the reason the tide is near ebb when the moon is directly above. The high tides, then, are caused by the waters displaced by the moon. It is also why, in Ellers's study of the jumping clams, they did not jump in the aquarium in the two hours around low tide even without a change in water level or other stimuli because they are responding to the difference in pressure of space-time compression, not the tide itself. With the increased compression of space-time on each other between two orbital partners pushing against each other's surfaces, even when there are no oceans to reveal this action, it will cause the orbital partner to synchronize their movements rotating in opposite directions, as we see, between the earth and its moon. This same effect between the sun and Venus is causing Venus to slow and reverse its direction of rotation.

Because the earth drags space-time on its rotation, we see space-time slowing the rotation of the earth. While this slowing would go unnoted, but for our sophisticated timing devices, it does accumulate notably over the earth's life span.

I can understand how some folks might have difficulty understanding how planets seem to move with such ease through space-time, yet it is the compression (bending of space-time) that creates the orbit. If I may return to our water analogy, we swim easily through the water, but when we are in a current, the current is carrying us along, although we are still swimming with ease through the water.

On our next stop, we decided if space-time is a physical property of this universe we need to define it a bit more. Since we can see through space-time as we see through air, we decided they must be like tiny bubbles.

Since space denotes a void, or emptiness, we decide to emphasize the time element of space-time and named the tiny bubbles tymeatrons. The tymeatron is flexible (compressible), has a high degree of exclusivity to matter (matter pushes into space-time), but a low degree of friction as the rotation of the earth shifts time but a small degree. We needed to decide how energy would travel through the tymeatrons, jumping from one to the next or slipping in between them. Because of the high degree of exclusivity to matter, we would expect the tymeatron and energy to have a high degree of exclusivity between them since energy is just another form of matter. While I'm sure in the future the tymeatron will be better described and undergo much further testing, we do not have to go off in search of some elusive new particle, as it was observed several decades ago. Film was placed in a box with two covered slots. When the cover was opened on either slot, the film was exposed, only beneath that slot. The light photons were simply pushing the tymeatrons out of the way and exposing the film directly behind that slot. But when both slots are opened at the same time, the film is exposed, not only beneath the slots but in a wide area. We now understand that since the light photons are pushing the tymeatrons away from both slots, whichever photon has the greatest energy forces the tymeatrons in front of the lower energy photon and deflects it to another part of the film.

In discussing time and gravity, it is easier to see if we are able to use analogies of the familiar. For time, we use the water analogy. Because we place ourselves in the position of importance, we consider time to be like a river flowing past us, whereas, time is more like an ocean and it is we who are passing through it. Large rotating masses drag time with them. This dragging creates currents in time effecting orbital planes. While no analogy works perfectly, it does help with general understanding, and the time-water analogy is helpful. For gravity, a looser analogy is a balloon. Think of a balloon. As we blow it up we are adding pressure and the balloon expands. Like most things, the balloon is a balance between the pressure inside the balloon and the pressure outside the balloon. If we add more pressure inside the balloon, the greater pressure expands the balloon. In a like manner, the greater the mass of an object in space, the higher the pressure toward space-time, and the greater the mass, the more it compresses space-time (the greater the curvature of space-time) and the greater the gravity. The worst part of this analogy is that the balloon has a membrane, while in gravity, there is no membrane as such, just the high degree exclusivity between matter pushing into space-time backed by a universe of space-time. As for humans, as the mass of the earth pushes into space-time and space-time pushes back against the earth, we are in the balance of force we call gravity. Unfortunately, the balloon analogy is the best I can think of at this time, with the internal pressure in the balloon pushing against our atmosphere being similar to the force of a massive object in space pushing into space-time.

Our next stop was how our solar system came into being. I think one of the big reasons astrophysicists are so reluctant to give up the pull theory of gravity even a hundred years after the release of *General Relativity* is that without the attraction between masses there is not, to them, any viable explanation for how this universe could come into existence. Of course, I do not have all the answers, but in science, there is one answer we have to use often, but reluctantly: "We don't know yet." We are always working toward knowing more, to answering the questions we have. Even having a question is a form of knowledge, for there is much that we don't even know that we don't know, so we can't even ask a question. But we digress from our stop here on the origin of our solar system.

The current theory is that a cloud of stardust, for some reason, collapsed (with the pull of gravity) into a hydrogen center, and when the pressure and temperature raised high enough, our sun started its fusion process (began burning). Meanwhile, the other elements balled up into the planets and moons. This is so wrong in so many ways. First of all, even strong pull gravity advocates realize we are in an expanding universe. Clouds of stardust are dispersing, not condensing. This scenario would be, unlikely even in a static universe, only theoretically possible in a contracting universe. Second, if pull gravity were to act on a condensing cloud of stardust, the heavier elements have greater mass and would pull toward the center and we would end up with a sun made of rock which does not burn well (no fusion).

We like beginnings and endings, but solar systems are not created; they evolve. Stars and planets do not have a common starting time, or a common beginning. First, stars start out as a burning ball of hydrogen (fusion). It is this fusion process that creates the higher elements from which solar systems are made. When stars mature, they may die with an explosion. This explosion does not form a cloud of stardust; it throws out chunks of all sizes. When these remnants of the star are caught up in the gravity field of another star, they become the planets and moon in that solar system. The smaller debris is soon incorporated into planets and moons through orbital intersecting. It could be that our solar system was primarily the result of an orbital partner of our sun . . . or perhaps not. It is a question which we may be able to answer within our children's lifetime.

In our next step, we step out from our solar system to the known universe. There are as many creation stories as there are religions, mostly anthropocentric in tone. Science, when it adheres to the law of parsimony, which is to use the fewest number of assumptions to cover all the known facts, can do but little better than most religions on this subject since we have so few facts and have to make so many assumptions. The major differences between science and religion is, in this area, no matter what theory science produces religion adds, "Yes, because God made it so" and the fact that science is easier to change when new facts are revealed, whereas religions are reluctant to incorporate change.

The current most popular scientific theory of the origin of the universe is the "big bang theory." This theory holds that in an instant, all matter was created in an intense hot burst of energy. After this creation, although expanding, it was still in a "hot dense state," which allows the stars to form from this dense cloud of matter and that gravity would pull these stars together to form galaxies. I have a lot of difficulty understanding how the universe has the energy to keep expanding, yet was able to condense these stars, or how galaxies could be pulled together from gravity while the universe was still young, but this pulling gravity did not simply collapse the whole universe. The answer given is there was just enough gravitational pull to create these stars and galaxies, but not quite enough to collapse the whole. The perfect balance, it seems. If we remove the pull theory of gravity as with relativity, then the whole big bang theory, as it now exists, becomes impossible.

Fortunately, with the discovery of black holes, we are

able to propose a new idea for the formation of stars and galaxies. First, we had to evoke a new principle, CDA (Cockroach Discovery Assumption), which states that a new discovery will be considered the new normal until proven otherwise. When a black hole was found at the center of a galaxy, it should be assumed there is a black hole in the center of all organized galaxies until proven otherwise. In the pull theory of gravity, this intense mass at the center of the galaxy would be pulling with such a force it would be holding the galaxy together, pulling some of the innermost stars into this hungry graveyard of stars. However, in the push theory of gravity, this enormous mass distorts space-time enough to create bending (compression) and a current in space-time large enough to encompass its whole galaxy. Further, the recent discovery that star speeds in a galaxy are the same, and the "HOT dog" galaxies strongly suggest that black holes are more than simply the center of galaxies; rather, they are actually the parents of stars, not the graveyard. In turn, this suggests that the building blocks of this universe are not stars, but black holes. The famous Penzias/Wilson heat signature does support a high-temperature event in the early universe, but I think we do not yet understand enough of about this universe to make much more than a guess as to the cause of this abnormality.

Let us take a moment to discuss the basic building blocks of an atom: the proton, which has a positive charge and is positioned in the nucleus, which is the bulk of the atom's mass, the electron, which has a negative charge, a small mass, and orbits around the nucleus, and the neutron, which is a proton that has been entered by an electron so that its charge is canceled out (or neither positive nor negatively charged) and has the weight of the combined proton and electron and is positioned in the nucleus. In the higher elements, neutrons maintain stability in the nucleus, which allows the positively charged protons to exist in the nucleus, although the protons are repellant to each other. In the nucleus of each atom, a certain range of neutrons will maintain stability in that atom. Too many or too few neutrons in a specific atom and it will be unstable and decay to a stable condition through an energy release.

So black holes are the earliest known building blocks of this universe. Some black holes erupted in a shower of stars creating galaxies. Within these galaxies we have the stars; dense masses of hydrogen, through nuclear fusion, which create the higher elements. At the end of a large star's life, these elements are distributed in various size chunks through the nova process to the surrounding area of the galactic space-time disruption or current where they are sometimes caught up in the gravity (bent space-time) of another star where they become planets. Also in the galaxies we have neutron stars. The current theory is that they are burned-out stars that have undergone a nova and blown out the heavier elements, and then collapsed back as neutron stars. As with the other theories of blowing away the heavier materials only to have the lighter material collapse back upon themselves, I find this theory wanting in logical consequence. I think it more likely that these neutron stars were part of the eruption process of the black hole and either too massive or still in the process of flowering (decompressing) into a hydrogen-fueled burning star. The fact that a neutron is just a compressed hydrogen atom is just too obvious to ignore. Since Edwin Hubble's work with red shift early in the twentieth century, we have known the universe is getting bigger. A more accurate description of this process is decompressing, rather than expanding.

I want to make a quick note here that the formation of galaxies appears to occur in two different ways, one way being an equatorial eruption to the formation of "HOT dog" galaxies, to evolve into spiral galaxies, and the second way being erupting out of the black hole poles or axis into barred galaxies. This too leads to more questions if this is correct that galaxies are created in different ways. Then do black holes themselves differ in composition? Since black holes appear to be spinning, is there an internal composition of subnucleonic particles in such agitation as to give black holes the effect of a dog chasing its tail? One of the greatest mysteries of this universe to me is why it is so dynamic and so many celestial occupants are spinning. Are the black holes the original spin?

Although we do not know a great deal about this universe, we should be able to agree that it is evolving. At this point, we have only a very limited knowledge of this process. Here again, I must take the uncomfortable position of questioning one of Einstein's ideas . . . that matter cannot be created or destroyed. I think we simply do not know enough about the evolution of this universe to make such a statement. Conservation of energy appears valid in local situations, but should not be assumed for the universe as a whole until we have a more complete understanding of it.

While we have provided a viable alternative to the current view of the evolution of the known universe, there are many questions we didn't tackle, because I don't know the answer, and it is our nature to prefer answers rather than questions. Nonetheless, let us pose a few of the questions.

If black holes are the first known building blocks of this universe, where did they come from? Are there other energy pools that we are completely unaware of that parented black holes, or was there a creation event as described in the big bang theory that created a black hole which later broke apart? The known universe is expanding (decompressing) and cooling. Is this decompression enough to trigger a black hole erupting into a galaxy, or are black holes undergoing a maturing process that is not yet known? Did the tymeatron originate with the beginning of this universe, or was it there prior to the beginning of the known universe? Is time the mysterious dark energy (dark matter)? Does time then have a unique parent other than black holes?

Last, I feel it necessary to evoke CDA in a couple of places where our anthropocentricity may have gotten in the way. First, is there other life in the known universe? Since there is an abundance of life on our planet, we must by right assume there is life throughout not only the known universe, but right here in the Milky Way galaxy where we reside. We don't even have a good estimate of how many stars are in this galaxy—something over 100 million. This is our own neighborhood, and that's as close as we can get! We have a lot to learn. Look around at the diversity of life here on earth and realize that life from another evolutionary scenario is likely to be very different from this planet's populous. So, yes . . . There is life out there.

As overwhelming as the size of the known universe is to us tiny creatures, and as little as we know about it, we still have to use the CDA and assume there are other universes, or that this universe is simply part of a larger system until proven otherwise. Now, let us turn our attention to the future.

Chapter 8
FUTURE

THIS LITTLE BOOK should make no difference in the daily routine of the vast majority of the world's population. Four hundred years ago it mattered little to the farmer or shopkeeper if the sunrise was the result of the sun orbiting the earth or the rotation of the earth on its axis. In a like manner, it matters little to us doing our sit-ups if the force we call gravity is the earth pulling on us or space-time pushing on us. Yet, without a broader understanding of our place in the universe, advances in science which affect many of us would not be possible. The GPS (Global Positioning System) which directs you to the correct road or guides the drone dropping bombs on you would not have been possible without the development of extremely accurate clocks, satellites, and compensation for the shift in time caused by the rotation of the earth. None of this is possible without the founding principles revealed by the theoretical work of some of the great names in science and the tireless work of the many unnamed engineers who created the practical application. I have no doubt that understanding time is a physical property will have practical consequences, but I shall only hazard guessing at a few possibilities.

This universe, of which we are but an insignificant part, begins with what is almost a contradiction: to the best of our knowledge at this time, it is decompressing, expanding, and unbound (not known to be bound), but within the universe, it is a finite interactive system. This universe is an ocean of time inhabited by black holes, galaxies, nuclear fusion balls we call stars or suns, and the resultants of their generation: planets, moons, and a wide variety of space debris. We have gleaned but very limited knowledge, and undoubtedly, there is much we cannot yet conceive. Because we see, as Einstein phrased it, large objects in space bending space-time and large rotating masses in space dragging time with them, and the understanding of the interaction between energy (photons of light) and the tymeatron in the two-slot box with film, we know we can deal with time mechanically. Dealing with time mechanically may sound exciting and give some readers a science fiction vision of living for hundreds of years . . . or eternally. Before we start talking about time-altering mechanical devices, let's call them time chambers, let us first remember that we are evolved of this earth; we are evolved passing through time at the most consistent speed of this earth. We are not equipped to physically or mentally to adjust to dramatic changes in a time-altered state. If our environment was too slow, we would be bored, or too fast we could not react to change adequately. If we had a time-altered chamber which was interacting with time at a slower rate, to those on the outside peering into the chamber, everything would be in slow motion. To those inside the chamber peering out, things would be flying by at an incomprehensible speed. Although those in the chamber might technically live twice as long as those outside the chamber, they would accomplish no more nor feel any longevity, having lived at slower time. There are, however, some conditions in which time altering might be beneficial, for example, medical procedures or space travel.

We all understand how our relative speed alters our passing through time. What else might alter passing through time? I think pressure (density of tymeatrons) may well alter the passing through time. Experiments in this area will be relatively simple to perform.

Since we see light pushing tymeatrons aside in the film exposure box, it appears we may be able to create a hole in time using higher energy beams. Einstein always referred to the speed of light in a vacuum; it would be more interesting and accurate for a base line to find the speed of light outside of time.

We can construct an easy variation to the double-slot box film exposure box to gain more insight into the tymeatron by adding a movable partition between the slots to separate the two chambers. We can vary this idea further by making a variety of positions available for the partition so it can be installed closer to one or the other slots to see if we can bounce the tymeatrons off a near wall and disrupt the light photons' path to the film. We can also have an adjustable aperture in the partition to limited effects of tymeatrons being pushed into each other's beams (pathways to the film). With this enhanced double-slot film exposure box, we may get a better understanding into "the dual nature of light, photons, and wave." It should now be understood that light is a photon of energy, and that the photon's travel through space-time (the wave motion) is exhibiting currents and movements of space-time (the physical properties of time).

I have been constantly using the closest analogy of space-time to the ocean with its expanse, currents, and waves. Perhaps in dealing with it mechanically, we may be able to use it in space travel the way we use water in jet boats.

As we grow accustomed to thinking about time as a physical property, we will find innovative and productive ways to utilize this simple fact.

Chapter 9
FINAL THOUGHTS

IN RETROSPECT, AFTER having reviewed the foregoing, so much of it seems obvious, almost self-evident; it is hard to say where I may have gone awry. I hope I have offered you a stepping-stone, not handed you a stone tablet. In dealing with the vast unknown, sometimes any idea is better than no idea as long as it is understood to be flexible and mutable . . . just a step on a long road, not a destination.

Gravity: to teach the two theories of gravity, that is, attraction between matter and relativity's curving of space-time, shows the same level of scientific malfeasance that teaches evolution and creationism together. The physics of this universe is the same here on earth as it is in the most distant galaxy. If one is to use the prerelativity concept of gravity as an attraction between masses, then they need to prove up. They have the space station in orbit.

It should be easy enough to set up an experiment there to prove the attraction between masses. Until then, we'll let astrophysicists puzzle over why gravity is not slowing down the expansion of the universe while we accept what Einstein told us almost a century ago, that there ain't no gravity (no attractive force between masses).

Orbits are one of the great balancing acts in this universe. Einstein explained them as masses falling in a straight line through curved space, but it is not quite so simple in this dynamic universe. In our solar system, which we know best, we have our moon orbiting earth, which is orbiting the sun, which is orbiting the black hole in the center of our galaxy. Because of the slight resistance provided by space-time orbits naturally decay, that is, the satellite declines toward the orbital partner, but because of the expansion of the universe, inherently decaying orbits are balanced for a longer interval, and at times the decline is even reversed. Thanks to the decompressing universe, our moon is gradually slipping away from us, and likewise, the earth is gradually slipping away from the sun. If not for the expanding universe, the moon would likely eventually plunge into the earth like other satellites, and the earth slowing, would likely plunge into the sun as we see stars near the black hole at the center of this galaxy plunging into it. The orbit of the earth around the sun is further complicated, in a minor way, by the sun's constant expenditure of energy (mass) and the earth's accumulation of mass through photosynthesis. The great cosmic juggling act which has allowed life on earth to flourish will end one day. Some distant day the final curtain will fall on what could have been a great run, but sadly, it appears likely that the show will close early.

Chapter 9
TIME (aside)

OUR DISCUSSION OF time for the purpose of this piece is to try to elaborate and explain that time is relative to the speed which we pass through the ocean of time in this universe (and it may vary in other conditions). Not only is time relative, but the measure of time is also arbitrary and random. Calendars here on earth vary between cultures. In Western cultures, our calendars are now fixed at 12 months of approximately 30 days. One culture in the pre-Columbian Americas had a 12-month, 30-day calendar beginning at the winter solstice with the 5 left-over days simply assigned to oblivion at the end of each year. In the Western culture, the annual 12 months of the moon were likely juggled to fit the full year. During the Roman Empire, Julius Caesar named July after himself and stole a day out of February to give his month 31 days. Not to be outdone, his successor, Caesar Augustus, named

August after himself and also stole a day out of February to give his month 31 days. It is likely that the calendar also started on the winter solstice but slipped with the extra quarter day per year into its current position. We now compensate for the extra quarter of a day per year with leap year every 4 years to give back 1 of the days to February, so badly bullied by the Caesars, and skipping leap year once every 400 years to adjust for long-term imperfections in trying to measure the earth's travels around the sun. In the Western world, this has become ingrained and standardized. It is our measure of history and utilized in most of science and mysticism. It is also likely the religious celebration of Christ's birthday (Christ being a symbol of renewal or new beginnings) was also timed to the winter solstice but slipped with the imperfections of the calendar.

In the older culture of China and parts of Asia, a variety of lunar calendars are still very much relevant. Many holidays and celebrations are timed by the lunar year. I do not know if the beginning of the lunar year was ever timed to coincide with the winter solstice. I think it more likely that the beginning of the year was timed more toward the beginning of spring. With lunar and solar sequences so misaligned between the two, the start of the lunar year varies against the solar calendar with occasionally an extra month being added to the lunar year to help correct the earth's rotation around the sun. There are still many people around this world who put great faith in the lunar calendar for predictions of which year will be fortuitous, which people are compatible with each other, etc.

Almost every culture that is known to us has had some variation of a calendar. With all the variation in the calendar between cultures, we get a small idea of how arbitrary and random measuring time can be. The clock, on the other hand, seems to have been accepted across all cultures. While this is handy in our time-driven society, it lends to the false impression that time is absolute rather than relative. We make note here that the measure of time is also somewhat arbitrary and random. The Chinese navigators of the early fifteenth century used a 10-hour clock, each of those hours being 2.4 of our hours. We are stuck with this base 10 counting system, most likely because of our 10 fingers, which likely was the earliest counting system. Ask a small child how old he is and he will hold up three or four fingers and tell you, "This many." Our mathematics would have been more logical had we had eight fingers. Nonetheless, we have a 10-based system, so it is surprising to me that we have a 24-hour clock. Will Durant seems to think 12 may have been used for its easy proportioning, that is, by thirds, quarters, or halves giving the modern clock two twelves. Likely, the sundial was one of the first clocks, and noon, when the sun was directly overhead, used as the major reference point. The day was divided in half; that is, half-light or daytime. The other, half-dark or nighttime, so we have two equal periods of 12 hours, with noon being the reference point instead of sunrise/sunset the two major time periods being the start. The new day ended up beginning in the dark, uncharted night at midnight. It is still surprising to me that we ended up with two 12-hour periods rather than two 10-hour periods or a 20-hour clock given our 10-based counting system, with perhaps 100 minutes in each hour and 50 seconds in each minute. We might have started the clock at sunrise on the equinox when the length of light and dark are about equal, rather than the noon/midnight clock we all use now.

If we were residents of Mars, the length of the day is similar to ours (about 26 of our hours), but takes nearly twice as long to orbit the sun or almost 2 of Earth's years. On Venus, on the other hand, the day or rotation on its axis is longer than its year or orbit around the sun. We can only wonder if life were able to evolve on either of our neighboring planets how their clocks and calendars would look. We can only guess how the more knowledgeable denizens of this universe are measuring their passage through time.

It is only from the height of our pretentious anthropocentricism that we are able to try to cast this universe in our concept of time as though this universe was our universe, instead of us being but a tiny speck in its unfolded grandeur, and yet we have no choice but to try to unravel this great mystery in the terms that we comprehend.

PART III:

AFTERWORD

POSTSCRIPT:
THE HUBBLE
TELESCOPE

MANKIND, AS FAR back as we can trace ourselves, have always used a great deal of symbolism, some of it simple and obvious as equating the plow to the penis and the fertility of the earth to the fertility of women, and other symbols more complex and ritualized in mysticism and religion. The art and literature of all cultures is rife with symbols. In the early 1900s, Freud, Jung, and the introspective psychologists worked to understand how the mind perceived and understood symbols.

When we read Melville's *Moby-Dick*, we can read a simple story of a whaling voyage, or do we, as many scholars still do, try to unravel deeper meanings and symbols. In the epilogue, we find Ishmael drawn into the whirlpool of the sinking *Pequod* until finally, at the center of the vortex, Queequeg's coffin pops to the surface and floats to him so he is able to hang on until he is rescued later by the *Rachel*, whose captain is looking for his lost son. In this simple last half page of *Moby-Dick*, we can see the deeply religious Melville's summation of life. Ishmael is drawn slowly around the larger outer rings of the whirlpool to the faster, smaller inner rings of the vortex. It is like life seems to us as the years get shorter as we get older until, at last, we reach the coffin in the center of the vortex, and it raises us up to be saved by "The Father" looking for his lost son.

Today, in our market-driven economy, we have seen symbolism turned to more purposeful uses from the happy, brightly colored animals and product placement to sell cereal to children, to the less sophisticated use of pretty ladies to sell just about anything to men as if the use of a particular product will heighten his appeal to these pretty ladies.

To me, the Hubble Telescope has become a symbol, a pinnacle of technical achievement. The creation of this magnificent machine and its launch into orbit is an unparalleled accomplishment. Some may think that the imperfect resolution of the original machine detracts from its amazing story, but I think it only heightens the greatness of this story that the brilliant women and men in this program were able to design a corrective component and the space shuttle program was to deliver and install it. Thanks to this marvelous machine, we common folks are able to peer into the untold realms of this great unfolding universe, to see with our own eyes wonders so far beyond our imagination we could not even dream of them. In one of the space shuttle's last missions before they were forced into early retirement, the battery pack in the Hubble was replaced by popular demand from the public and our eye on the universe had its life extended a few more years, yet there is so much left unseen in the universe and so little battery life left. It is not that our scientific discovery of the universe has stopped. We are used to technology being rapidly replaced by something new and improved, but there really is no newer, better Hubble. When this spectacular machine goes dark, it will be, to me, a symbol of the darkness descending across our little planet.

There is a darkness spreading around our globe, and it is a blighting of the mind and spirit of mankind. We see it in using religion to justify throwing acid on girls for attending school, or shooting girls for wanting an education. We see it in India using religion to depreciate, deprive, and plunder "the lower casts." It is always easier to see the weeds in our neighbor's backyard than to see those in our own front yard. In the USA, we are as guilty as any other nation of religious excesses, of using religion to justify violating logic and common sense. I have often heard fundamentalists use the expression, "It's all in the Bible." No. There are no cell phones, no semiconductors, no telescopes, no televisions, no Internet, no rockets, no nuclear weapons, not even a gun in the Bible. But if there is a God, he gave mankind the ability to create these things through logic, critical reason, rationality, and deduction. These things were invented and created by men, not prayed into existence.

Frederick Douglass, in his autobiographic narrative of his life as a slave, recounts how his mistress secretly had been teaching him the alphabet and a few small words and the scene that ensued when her husband discovered them. The master raged that it was unlawful, as well as unsafe, to teach a slave to read. Mr. Auld said, ". . . Learning would spoil the best nigger in the world. . . . It would forever unfit him to be a slave." For Frederick, still only a lad of eight, it was an epiphany. Even as young as he was, he now had the answers to the question that nagged him: how the white man enslaved the black man and his personal path to freedom. Mrs. Auld joined with her husband and with the enthusiasm of the recently converted, her vigilance of Frederick to protect him from learning was intense, but Frederick now had the insatiable desire to learn that could not be stopped.

Frederick's narrative, while written as a testimonial for the abolition of slavery, abounds with psychological insights as vital as ever. Half a century before Sigmund Freud began to liberate psychology from the realm of witchcraft and sorcery, Frederick Douglass had remarkable insight into the human mind, much of which is still relevant today. His condemnation of the slave masters who often prayed the loudest but whipped their slaves most frequently and the hardest, caused Frederick to write an addendum to explain that he himself was also a Christian, but a Christian who believed in Christ's teachings of charity and goodwill toward his fellow man, not a religion that promotes slavery and abuse of our fellow man. It is a prime example of those who say, "It's all in the Bible" . . . They are essentially saying anyone can justify anything no matter how horrifying or illogical. So here we are today, using our religion as a justification to throw acid on girls for seeking education, raping girls in India to help God punish them for the imperfections in their past lives which caused them to be born in the lower station in life as a female, or using the power of religion to cast doubt on logic and science.

The Bible and other religious writings are full of wisdom. Religion does not have to be an obstacle to education. Many people in the nineteenth-century United States grew up on farms where there was little time for education and learned to read with the only book they possessed: the Bible. But the Bible whetted their appetite for more learning. It served as a gateway to education, not an end. Sadly, if mankind spent even half the time studying science and philosophy as they spend memorizing Bible verses, it would be a boon, both to the earth, and all its inhabitants. But, alas, we are what we are, and we will go on beating each other over the head with our holy books until, as Madame Jullien said, "But all that is leading us towards a catastrophe which will cause the friends of humanity to shudder; for, it will rain blood. I do not exaggerate." Even as the darkness descends, dimming lights of learning, it brings forth the black humor and obvious irony that those who "are full of passionate intensity," beating each other with their holy books, would be on the other side still blindly following, but for randomness of their place of birth.

Having reread the above paragraphs, I feel the need to clarify the role between religion and education. Our human personality requires most of us to feel we are included in some "special group." This special group, or "tribe," is often acquired simply by birth. We are born in a particular town and state and country and heritage and often religion. We often acquire other tribal affiliations through schools, clubs, social organizations, sports clubs, or gangs. Most of these tribal affiliations require a form of unquestioning loyalty to the group and a depreciation of the other groups to make the assumption that we are better than others. Religion has always had a powerful place in the human personality and likely always will. It has always taken a "leap of faith" to accept any religion. The conflict between science and religion always occurs when facts are ignored to sustain that faith (belief-biased effect), as mentioned before, the religious insistence on the geocentric view of the world long after science provided overwhelming evidence of the heliocentric view.

As a child, I used the words *ignorant* and *stupid* interchangeably without distinction, but there is an enormous difference. *Ignorant* is lacking education; *stupid* is foolish or senseless. Monkey see, monkey do, is often used as an insult suggesting mimicry, mindless imitation. Actually, when we see someone doing something successfully and we wish to accomplish that same goal, then we do as they are doing. That is part of how we learn, how we are educated, how we doff our ignorance. But when monkey see and monkey doesn't do, well, that's just plain stupid. In the USA, we see the education system we have costing more, but being far less effective than other educational systems, and yet, we don't implement those systems—well, that is just stupid. It is the same for the medical system. It's far more expensive, far less effective—and yes, that's just stupid. The list goes on for the USA of things we are paying more for and getting poorer results than other countries, and that is just stupid . . . frustratingly stupid! This is not the forum to address these issues; I simply use this to point out the caustic erosion learning in the U.S. public where we monkeys see, but we monkeys don't do.

Paleontologists have long assumed brain size (cranial capacity) during human evolution equated to intelligence. Given the fossil record and artifacts used to make these estimates, we might accept them as generally correct and too, the fact that we have no other way of evaluating this thinly documented part of evolution, we've little choice but to accept them, albeit with reservation, for we see by our limited understanding of savant syndrome that our brain has a much greater capacity than we are able to reach or unlock. What we do know is the cranial capacity of modern man is decreasing. Our brains are getting smaller: decreasing from a median average of 1,500 cc about 15,000 years ago to about 1,350 cc today. The decrease roughly corresponds to the Neolithic Revolution, the shift from hunter-gatherers to agrarian cultures and domestication of food sources. One theory holds that as we assumed the specialized roles in agrarian cultures and population centers, our narrower range of skills required less cognitive ability, hence, the decrease in brain size. Another theory holds that humans became less aggressive during this time. While I'm not at all sure of this in the strictest sense, I do think that larger groups afforded greater protection from wild animals as well as predators of our own species, thus, lowering our cognitive requirements. Memory requirements were greatly lowered with the invention of writing, and even more with the printing press, and at this point, with Internet access to all sorts of information, we have far less need of memory. The last theory of lower brain capacity that I'm aware of is that we are simply getting dumber. None of these theories precludes any of the others.

Modern history seems to run in cycles, from ages of learning to ages of anti-intellectualism; from ages of enlightenment to darkness. We put men on the moon and brought them home. We put the Hubble in orbit and repaired it and renew its batteries. We still have the intellect to go forward, moving pieces of the unknown into the known, into our common knowledge, of raising ourselves and all mankind to greater heights. But we won't! We spend our curiosity following the exploits of celebrities not our scientists. Much of the drudgery of our lives has been removed with labor and time-saving machinery and devices which free us for self-improvement or higher pursuits, but that time is simply squandered on the shallowest of entertainment. The bread and circus of Rome are now the subsidized food and dumbed down and vulgar digital entertainment: the homeostasis of a devolving psyche. Evolution rewards intelligence; it deals harshly with stupidity. In past periods of darkness there were still pockets of light, but in our globalized world, one wonders where these beacons of light will be. Meanwhile, the footprints of our explorers on the moon are filling with dust, not followers, and the marvelous Hubble telescope draws ever closer to a premature demise.